SKILLED AND VOCATIONAL TRADES

BECOME A CONSTRUCTION AND BUILDING INSPECTOR

by Elizabeth Hobbs Voss

BrightPoint Press

San Diego, CA

BrightPoint Press

© 2024 BrightPoint Press
an imprint of ReferencePoint Press, Inc.
Printed in the United States

For more information, contact:
BrightPoint Press
PO Box 27779
San Diego, CA 92198
www.BrightPointPress.com

ALL RIGHTS RESERVED.

No part of this work covered by the copyright hereon may be reproduced or used in any form or by any means—graphic, electronic, or mechanical, including photocopying, recording, taping, web distribution, or information storage retrieval systems—without the written permission of the publisher.

LIBRARY OF CONGRESS CATALOGING-IN-PUBLICATION DATA

Names: Voss, Elizabeth Hobbs, author.
Title: Become a construction and building inspector / by Elizabeth Hobbs Voss.
Description: San Diego, CA: BrightPoint Press, [2024] | Series: Skilled and vocational trades | Includes bibliographical references and index. | Audience: Ages 13 | Audience: Grades 7-9
Identifiers: LCCN 2023009794 (print) | LCCN 2023009795 (eBook) | ISBN 9781678206826 (hardcover) | ISBN 9781678206833 (eBook)
Subjects: LCSH: Building inspection--Vocational guidance--Juvenile literature.
Classification: LCC TH439 .V67 2024 (print) | LCC TH439 (eBook) | DDC 690/.21--dc23/eng/20230321
LC record available at https://lccn.loc.gov/2023009794
LC eBook record available at https://lccn.loc.gov/2023009795

CONTENTS

AT A GLANCE	4
INTRODUCTION	6
A BUILDING INSPECTOR AS A SUPERHERO	
CHAPTER ONE	12
WHAT DOES A CONSTRUCTION AND BUILDING INSPECTOR DO?	
CHAPTER TWO	32
WHAT TRAINING DO CONSTRUCTION AND BUILDING INSPECTORS NEED?	
CHAPTER THREE	42
WHAT IS LIFE LIKE AS A CONSTRUCTION AND BUILDING INSPECTOR?	
CHAPTER FOUR	60
WHAT IS THE FUTURE FOR CONSTRUCTION AND BUILDING INSPECTORS?	
Glossary	74
Source Notes	75
For Further Research	76
Index	78
Image Credits	79
About the Author	80

- Construction and building inspectors make sure construction is done properly. It must follow building codes.

- States with a greater risk for natural disasters have stricter building codes.

- Inspectors usually specialize in a particular field. There are electrical, plumbing, and public works inspectors. There are also housing, fire prevention, and mechanical inspectors.

- Building inspectors need a high school diploma. Some jobs also require an associate's degree, a certificate from a trade school, a bachelor's degree, or a master's degree.

- Most states require a license or certification. Usually, this must be renewed regularly.

- Inspectors may work on construction sites. They also do paperwork and schedule inspections in field offices.

- Inspectors may use tools such as flashlights and infrared thermometers.

- Inspectors living in large cities may earn more money than those living in the suburbs.

- Organizations and internship programs are working to recruit more women and young people into the inspection field.

INTRODUCTION

A BUILDING INSPECTOR AS A SUPERHERO

Hanging from ropes like a mountain climber, April Turner skims across the high-rise. Turner is an assistant project manager. She inspects building exteriors and makes sure they are safe.

Turner is 188 feet (57 m) above the sidewalk. She slides past windows

Inspections of tall buildings and other large structures are sometimes done while suspended from ropes.

and balconies. She holds a tiny **mallet**

and camera. She takes pictures of cracks.

She feels for flaws. When she is finished,

she detaches herself. She walks inside

the apartment building. Then she takes an

elevator to the roof, where she reattaches herself to ropes. She drops down to look at another section.

In New York City, buildings taller than six stories must have the exterior inspected every five years. Rope-access inspections are becoming more popular. They replace swing stage scaffolds. This is when a platform is hung by cables. The platform holds workers and their tools. Using ropes is faster and less expensive.

There are many men in the construction field. However, more women have been interested in rope-access inspections.

Historically, building inspectors have most commonly been men, but more women are joining this field today.

Turner's rope-access training class was almost entirely female. Two men started out in the class, but one dropped out.

Amy DeLuca saw a photo of someone **rappelling** down the side of a building. She said to herself, "I want to do that."

She became a rope-access inspector in 2015. Xsusha Flandro is another woman who inspects tall buildings. She has examined about twenty buildings in New York. Flandro is usually the only female on a construction site. It's different on the ropes, where there are far more women.

WHAT IS A CONSTRUCTION AND BUILDING INSPECTOR?

Few construction and building inspectors hang from ropes. Most inspectors spend their time on construction sites. Besides buildings, inspectors may examine highways, bridges, and sewer systems.

Learning to work safely at great heights takes a lot of training.

If a building is under construction, inspectors may look at electrical and plumbing systems. Usually, inspectors have specialties. This is because there are so many things to examine. Construction and building inspectors are important. They make sure people can live and work in safe places.

CHAPTER ONE

WHAT DOES A CONSTRUCTION AND BUILDING INSPECTOR DO?

Construction and building inspectors protect public health. They make sure buildings and other structures are built safely and correctly. Structures must follow building codes, zoning rules, and architects' plans.

Clearly marked fire exits are commonly required by building codes.

BUILDING CODES

Building codes are rules created by national, state, and local governments. The codes have requirements for how structures need to be designed and built.

The codes address many safety concerns. For example, codes make sure buildings have safe wiring. They ensure that there are enough fire exits. They ban unsafe materials. For example, wood shingle roofs are not allowed in Los Angeles, California. Wildfires are common there, and wooden roofs can catch on fire easily. Inspectors make sure buildings are safe from fires.

Codes are constantly changing to keep up with new technology. For instance, codes are updated to include green energy technologies. This includes solar panels and battery storage. Updated codes can help buildings save energy. They also lower the buildings' greenhouse gas emissions. Greenhouse gases trap heat in the atmosphere. This leads to climate change. Updated codes help address threats from the changing climate.

Florida has one of the strongest building codes in the country. This is because

SAN FRANCISCO SEEKS TO LIMIT EARTHQUAKE DAMAGE

In San Francisco, California, earthquakes are common. The city's building code prevents major structural failure and loss of life. Wood frame buildings of three or more stories must be strengthened to make them more earthquake resistant. The code minimizes damage to buildings after earthquakes.

Some remodeling of historic buildings helps make the structures more resistant to earthquakes.

hurricanes are common in Florida. Making buildings tough can help them survive destructive natural disasters. This can also save a lot of money.

Natural disasters can be devastating. Between 1980 and 2022, the weather was responsible for $2.5 trillion in losses in the United States. Stronger buildings are damaged far less.

The Federal Emergency Management Agency (FEMA) develops hazard maps. These maps help people construct buildings away from dangerous areas. FEMA also recommends stronger building codes.

Sometimes older buildings must be brought up to current codes. Remodeled buildings must be inspected. This makes sure that they are safe and sturdy.
The government creates guidelines for preserving and restoring historic buildings.

ZONING AND BUILDING SPECIFICATIONS

Inspectors make sure structures follow zoning rules. These rules control what can or cannot be built in an area. There are different types of zoning districts. One zone may be residential. Another may be commercial. A third may be industrial. Zoning rules may control the sizes of lots and buildings. They can protect historic places. They may tell how much space there should be between buildings. Rules in a commercial district can require a certain number of parking spaces. Inspectors look

Residential zones include apartments, houses, and condos.

at plans for a building. They make sure buildings follow zoning rules.

New structures must follow contract specifications, or specs. Architects and designers usually write the specs. They describe how a building should be made.

Specs detail what materials should be used and how they should be installed.

THE INSPECTION PROCESS

In new construction and some remodeling projects, builders apply for permits. These are required before construction can begin. Inspectors review a builder's plans to make sure they follow codes. If they do, inspectors issue building permits. In the first stages of the project, inspectors examine materials to make sure they meet the specs. During construction, they do regular checks. They look for any violations of building codes.

Before construction even starts, inspectors may review the builder's plans to ensure they are up to code.

If inspectors find problems, they give violation notices. This means the builder must quickly correct the problem. The builder needs to make sure the project follows the codes.

Sometimes inspectors issue stop work orders. They do this if a construction site is unsafe. All work must stop except work required to make the site safe. A partial stop work order is when some work can continue while the problem is being fixed. When a stop work order is issued, inspectors contact the project executive or construction manager. They explain the necessary steps to correct the safety hazard. If the stop work order is violated, the government may issue fines. The stop work order is lifted after the inspector returns and finds the problems corrected.

Stop work orders protect workers and the public. They also protect structures from unsafe conditions.

Building inspectors keep records. These include records of permit applications, permits issued, and fees collected. Each inspection is noted. Notices and orders are also recorded.

HOW ARE HOME INSPECTORS DIFFERENT FROM BUILDING INSPECTORS?

Home inspectors are not building inspectors. They are usually hired by home buyers to look at homes the buyers are considering purchasing. Home inspectors do not look for code violations. Instead, they look for safety concerns or repair needs in houses.

Often, inspectors work alone. However, they may work in a team of specialists. They interact with building owners, contractors, and others. They schedule inspections daily. They know how to deal with conflict when people disagree with their evaluations.

NEW YORK CITY DOES A SAFETY SWEEP

In the first half of 2021, seven construction workers died in New York City. Most deaths were from falls. The city launched a strategy in June of that year. Inspectors went to the city's largest construction sites. They made sure workers were using safety harnesses and **fall arrest systems**.

New York City's many tall buildings can make building inspection a difficult and even dangerous job.

During the three-month campaign, inspectors issued more than 3,600 violations. They also issued 1,499 stop work orders. "The recent spate of construction worker deaths in our city

CONSTRUCTION AND BUILDING INSPECTOR SPECIALTIES

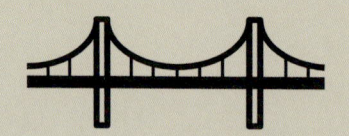

Public works inspectors
- Look at government projects, such as highways, bridges, and dams

Plumbing inspectors
- Check plumbing
- Look at sanitary systems
- Examine storm sewers
- Check water supplies

Housing inspectors
- Look at housing
- Check for safety violations
- Examine exterior property

Fire safety inspectors
- Work in fire departments
- Check buildings once a year
- Make sure fire safety practices are used

Construction and building inspectors have many specialties.

is tragic, senseless—and even worse, entirely avoidable," said New York Building Commissioner Melanie La Rocca.[1]

INSPECTORS USUALLY SPECIALIZE

In some towns, one person may inspect buildings. The building inspector can look at plumbing, mechanical issues, and electrical work. However, in larger cities, inspectors specialize. One may do public works. Another may handle housing. A third may focus on fire prevention. Inspectors are usually certified in one or more disciplines.

Pete Jackson is the chief electrical inspector for the city of Bakersfield, California. He said, "After 25 years as an electrician and electrical contractor, I needed a different challenge. Installing systems as quickly and cheaply as possible no longer interested me. Ensuring the correct application of code and resulting correct installation is a more interesting challenge."[2]

INSPECTOR SAVES CONTRACTOR MONEY

Michal Hofkin told a story of how he saved a contractor thousands of dollars. Hofkin is a code specialist for UL Solutions. UL Solutions is a global safety science company. The contractor was going to put in plastic wiring. However, an engineer said it violated the building code. Hofkin came

Inspectors and contractors may have to work together to resolve issues that come up during construction.

to the contractor's defense. He ended up in an important meeting with the general contractor, the building owner, and the architect. Hofkin explained what the code

29

said and what it meant. In the end, he was proven correct, and the project continued. Hofkin explained:

> *If the contractor had been forced to use a metal wire instead, it would have cost him tens of thousands of dollars. Inspectors are not the enemy. We are the front line for public safety. Contrary to popular perception and rumor, we are not there to cost you money, we are there to protect public safety. . . . Inspectors should be considered as assets, not liabilities.*[3]

Inspectors make sure buildings are safe for people to be in.

CHAPTER TWO

WHAT TRAINING DO CONSTRUCTION AND BUILDING INSPECTORS NEED?

Building inspectors need high school diplomas. They may also need an associate's degree or a certificate from a vocational school or community college.

A college student with an interest in architecture may choose to go into the inspection field.

Community colleges offer associate's degrees in trades. Vocational or trade schools are another choice. Some employers want employees to have bachelor's degrees. The degree may be in business, property management, or architecture.

CHOOSING THE RIGHT PROGRAM

Community college programs should include courses in algebra, geometry, and writing. These help inspectors read blueprints and write reports. Vocational schools prepare students to enter fields that require specialized training. Students earn certificates and graduate after two years. They do not take courses outside their fields of study.

All programs should prepare students to become certified, licensed inspectors.

Working with an experienced mentor can help future inspectors gain valuable real-world experience.

Programs should include field training.

They also should prepare students for national and state exams.

The best programs offer mentoring.

The mentors are experienced inspectors.

Students follow them on the job. They learn codes, techniques, and terminology. They pick up communication skills.

EXPERIENCE IN CONSTRUCTION TRADES

Employers like inspectors with a background in the construction trades. Construction trades include any job involving construction of buildings, roads, or bridges. Many inspectors have worked in trade jobs. They may have been plumbers, carpenters, or electricians.

Before Pete Jackson became an electrical inspector, he was an electrician and electrical contractor for

Someone with expertise in electrical work may be a great candidate to become an electrical inspector.

twenty-five years. Michal Hofkin owned and operated his own electrical contracting firm for ten years. He also worked as a building code instructor. Hofkin encourages teens who are interested in becoming inspectors. He tells them to first learn a trade and the building code. "Being an electrician and code instructor prepared me very well to be an electrical inspector," Hofkin said. "I was hired fifteen seconds into the phone call during which I called an inspection agency to see if a job was available for me."[4]

Training requirements vary by state and type of inspector. Generally, inspectors learn a lot on the job. In the beginning, they work with more experienced inspectors. They learn techniques and codes. They also learn about recordkeeping. Inspectors keep daily logs with photographs from inspections.

GETTING A LICENSE

Most states require a license or certification to become an inspector. Some states have their own programs. In other states, inspectors need certificates from associations. One of these groups is called the International Code Council (ICC). Another is the National Fire

PREPARING FOR LICENSING EXAMS

Some people take prep classes to prepare for the inspector licensing exams. However, courses can be expensive. Students should take practice exams if they are available. Most of the time, there is no limit to the number of times someone can take the exam.

Potential building inspectors must pass tests to prove their knowledge.

Protection Association. Getting certification can involve attending seminars. It also includes learning building codes and passing an exam. The exam tests knowledge of construction technology and inspection procedures. In Minnesota, the building inspector exam is called the

Certified Building Official Limited Exam. Aspiring inspectors take a five-day training course. They also must fulfill education and training requirements.

States may have different certification requirements. Candidates typically must pass an exam and purchase **liability insurance**. They need a minimum level of education and building inspection experience. In most states, building inspectors need to renew their licenses regularly. They also need to take more education courses. This gives them up-to-date knowledge about the field.

CHAPTER THREE

WHAT IS LIFE LIKE AS A CONSTRUCTION AND BUILDING INSPECTOR?

Inspectors spend most of their time on construction sites. That is where they do their inspections. They also work in field offices. This is where they review blueprints, do paperwork, and schedule inspections.

Much of an inspector's work is done at a construction site.

CAN THE WORK BE RISKY?

Working on construction sites can be hazardous. Inspectors may encounter toxic chemicals, poor air quality, and dust. They need to work around large construction equipment, which can be dangerous.

They are sometimes exposed to very hot or cold temperatures. They are often around noisy equipment.

All inspectors receive safety training on a regular basis. The training follows federal, state, and employer requirements. Code specialist Michal Hofkin says the best way

INSPECTOR GEAR

Building inspectors wear personal protective equipment (PPE). This includes coveralls, safety shoes, and goggles. If they work on roofs, they may bring tie-offs or lanyards. A lanyard is a flexible rope, wire rope, or strap. The inspector wears a harness. The lanyard is attached to the harness and anchored to the roof. This keeps inspectors safe.

inspectors can stay safe is to stay alert and avoid dangerous situations. If they see something hazardous, they should report it to the construction manager.

WORK HOURS

Most inspectors work full-time during regular business hours. However, their work schedules can be irregular. Schedules can change with the weather. Projects take differing amounts of time. Inspectors may work extra hours during busy seasons.

If an accident happens, inspectors must respond right away. They may need to work longer to complete reports

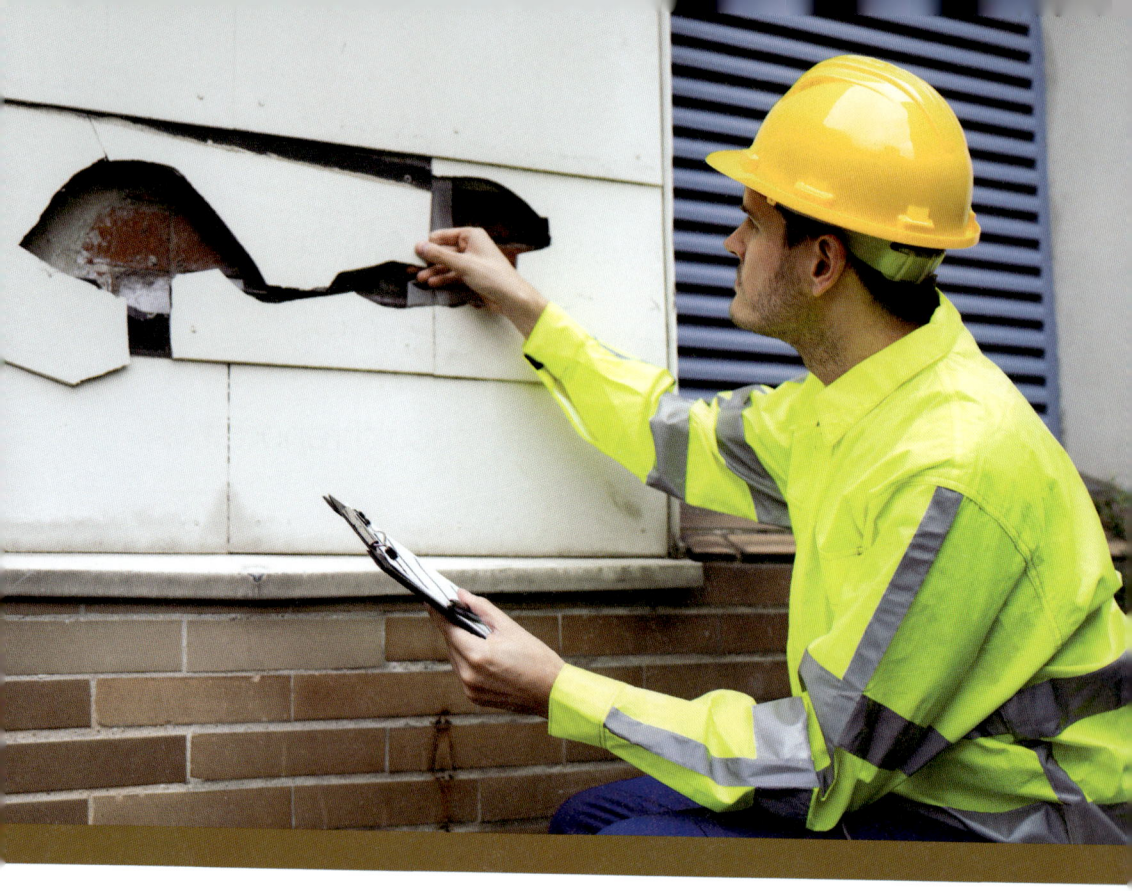

Inspectors may be called in when there is an accident or incident with a building.

about the accident. For example, Pete Jackson investigated a roof fire at a Target in Bakersfield, California. He made recommendations based on his investigation. As a result, new code and product requirements were written to prevent any future fires. Sometimes, inspectors need to work evenings and weekends. This is especially true for those who are self-employed.

About 38 percent of building inspectors work in private, for-profit industries. Another 41 percent work for state or local governments. Seventeen percent are self-employed. One percent work for the federal government. Finally, 2 percent work in not-for-profit industries.

IMPORTANT SKILLS NEEDED

A wide range of skills are used as a building inspector. It is important for those interested in the job to have these skills. Employers will be looking for them.

Both written and oral communication skills are important. Inspectors must keep logs of their inspections. They also need to write reports about their findings. Effective writers can clearly explain complex topics. Inspectors also talk to owners about their property's condition. They communicate about correcting any problems. Inspectors need to explain technical details in an easy-to-understand way.

Inspectors must be detail oriented. They work on large projects. They need to pay

Attention to detail and communication skills help inspectors identify problems and make sure they get fixed.

close attention and notice small details. For example, when examining a building's foundation, they might see cracks in the concrete. They need to make note of it and take necessary action.

Inspectors might need to enter tight spaces to make sure a building's structure is sound.

During field work, inspectors are always on their feet. They may have to climb and crawl though attics and other tight spaces. They should be physically fit.

Inspectors use problem-solving skills to identify and resolve issues. For instance, an inspector may notice that there's an electrical issue in one room but not in another. The inspector needs to investigate and find the source of the problem.

It is important for inspectors to manage time wisely. Inspectors may have several projects they are doing at once. Good time management can help them complete inspections. It can help them submit reports to their supervisors on time.

Inspectors also need customer service skills. They must answer questions and

Simple tools, such as flashlights, are crucial for many building inspectors.

address concerns. They need to provide

information to owners, contractors, and

subcontractors. These skills help them maintain good professional relationships with clients, coworkers, and employers.

WHAT TOOLS DO THEY USE?

Inspectors need many tools. A flashlight is a necessity. A small, pocket-sized one may help in tight spots. A telescoping mirror allows inspectors to see in difficult-to-reach areas. A telescoping ladder folds down to short lengths. It is easy to carry.

Inspectors wear tool belts or vests. These help them carry their tools, keeping their hands free. Vests have many pockets with secure closures.

Inspectors can use drones to see and photograph hard-to-reach areas.

Building inspectors carry electrical testers. These include AFCI or GFCI testers. They are used to test electrical outlets. Inspectors also carry voltage indicators.

These show if electrical current is flowing through wires. Infrared thermometers and cameras show heat changes over a large area. Unexpected heat changes can mean air leakage or heat loss. Inspectors may also carry gas detectors. These devices will find if there is any poisonous gas.

AN ELECTRICAL INSPECTOR'S DAY

Pete Jackson is a chief electrical inspector. His day begins at 6:30 a.m. He reviews the 150 to 200 inspections scheduled for the day. He makes sure the right inspector is assigned to each one. He decides what

The UL logo on electrical equipment shows that the product has met safety standards.

inspections he needs to do. This takes about two hours.

The rest of Jackson's day is divided between field inspections, phone calls, reviewing construction plans, and meetings. His work day ends at 4:00 p.m. Like most inspectors, he works a forty-hour week,

Monday through Friday. He is a member of several code-making panels for the International Association of Electrical Inspectors. These groups help put together the National Electrical Code and UL Solutions Product Standards.

The National Electrical Code is a US standard to safely install electrical wiring. Safe electrical wiring helps prevent fires and other accidents. UL Solutions is a global safety science company. It puts together standards for product safety. If a product has met a standard, it has passed tests that check for safety and quality.

SALARY

The median annual wage for construction and building inspectors is $61,640. That means half the workers earn more. Half the workers earn less. The lowest paid earn less than $38,110. The highest earn more than $100,520.

Salaries depend on experience, education, and location. Building inspectors with a bachelor's degree earn about $68,697. Those with only a high school education make about $60,002. Salaries are higher in some states and territories and lower in others. Inspectors have the highest

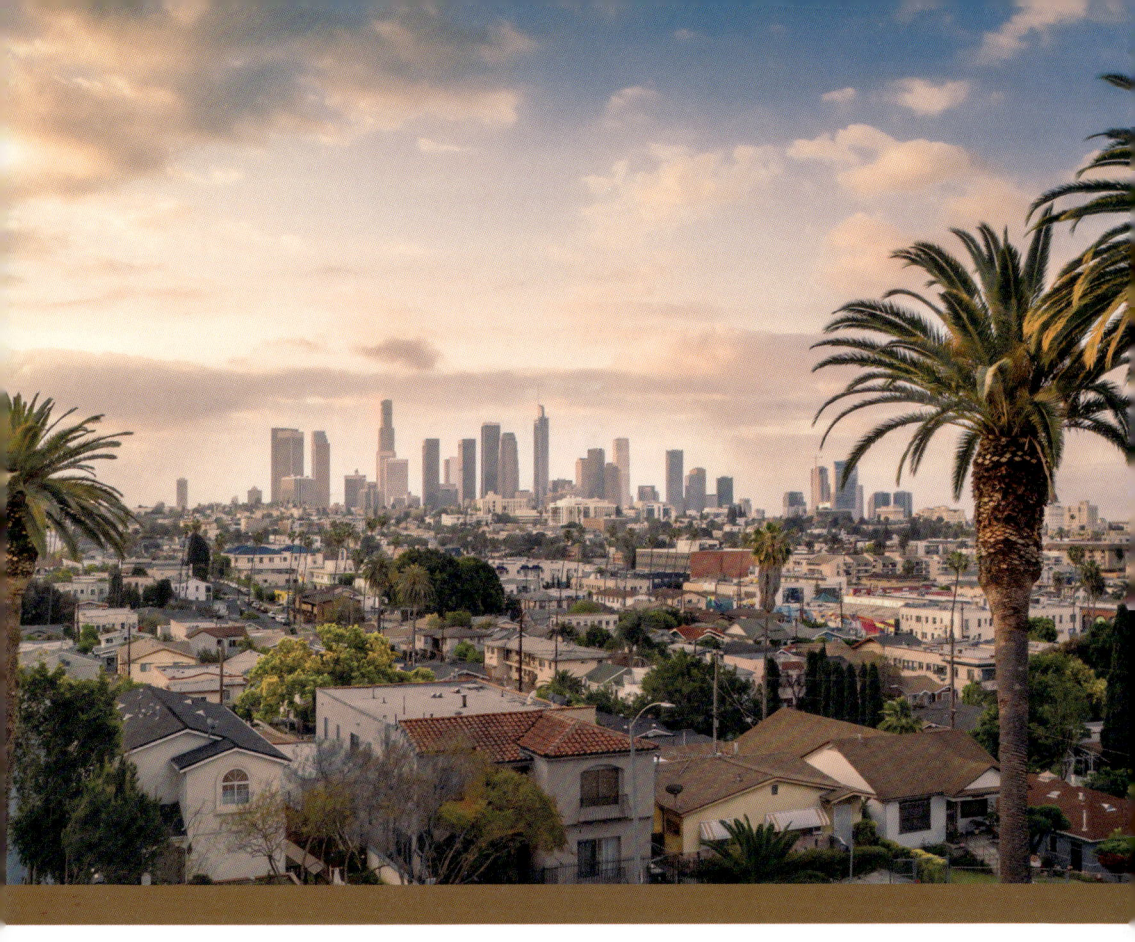

Inspectors living in California earn more than those living in Arkansas.

average salaries in California, Alaska, and Washington. The lowest salaries are in Arkansas. City inspectors tend to make more than rural inspectors. For example, the average salary in New York City is $70,258.

CHAPTER FOUR

WHAT IS THE FUTURE FOR CONSTRUCTION AND BUILDING INSPECTORS?

Employment in inspector jobs is expected to decline by 4 percent from 2021 to 2031. Still, about 14,800 openings are expected every year. These are to replace workers who leave

As older inspectors retire, the industry will need to train a new generation of these important workers.

the field. Some inspectors transfer to other occupations. Others retire or leave the workforce.

In recent years, there has been a demand for more people in construction. More high schools are bringing back classes that bring awareness to careers in this field. Mentorships and training programs are also helping to attract young people to become inspectors.

REACHING OUT TO YOUNG PEOPLE

In 2018, a building official in Wyoming, Jim Brown, started a shadow program for young professionals. The program offered $2,000 scholarships to mentees. The mentees shadowed the International Code Council (ICC) region president at an annual conference. The International Code Council is a global association for building safety professionals. The shadowing program ended successfully.

Some programs give high school students a chance to see what the building inspection field is like.

About 90 percent of the mentees had found career positions in their community. Participating in this type of program can give future inspectors a glimpse into the career.

Inspector Jason Pryor was also inspired to reach out to young people. He is a building inspector for the city of Gulfport, Mississippi. He started a mentorship program there. His goal is to launch a statewide high school mentorship program. "In the building code profession, forty is still young," he said. "The challenge is getting people to take the initiative to participate."[5]

The average age of male building inspectors is 48.7. For women, the average age is 46.2. Often, building inspectors start working in the construction trades. Both Michal Hofkin and Pete Jackson first worked as electricians and electrical contractors.

In Colorado, an ICC chapter started a technical training program in high schools. Training materials were provided. Local construction companies donated tools. Mentors helped students earn certifications. They helped review test questions. They answered questions about careers.

ONLINE PROGRAM ATTRACTS YOUNG PEOPLE

The International Code Council has a program called Safety 2.0. Its purpose is to attract young people to building safety professions. People learn online. They get training. They find mentors. They get certified. More than 12,000 people have participated.

The construction and building inspection industry is making an effort to bring more young people into the field.

High schools are becoming more interested in programs like this.

Hofkin has been a code instructor for thirty-four years. He agreed there are fewer students today than there were twenty years ago. "Our culture does not put the same value on going into the trades as it does on a college degree. . . . I have nothing against a college education, but it is not for everyone. For those who wish to work with their hands and their minds, trade school may be the better educational path," he said.[6]

Some states have internship programs for inspectors. A Florida building association launched its program in 2017. Candidates first pass an ICC exam. Then they work with a professional inspector for four years under a **provisional** license. They take a training course and finally pass a state exam to earn a license.

A new generation of inspectors is changing the demographic makeup of the industry.

DEMOGRAPHICS

Building inspection is a male-dominated field. The gender breakdown is 88 percent male and 12 percent female. Women account for only 10.9 percent of the construction industry's labor force.

They are outnumbered nine to one in construction jobs. Women's construction jobs tend to be in sales or office positions. However, there is a trend toward more women in the construction industry. The number of women in construction increased 54.7 percent from 802,000 in 2012 to 1,241,000 in 2021. Also, the number of women in construction management roles increased 101 percent between 2016 and 2021.

However, gender discrimination is a problem in the construction industry. Women are less likely to be hired. They are given fewer hours. They are three times more likely to be passed over for a promotion. Unfortunately, there are still **stereotypes** about women's abilities in the field.

The National Association of Women in Construction combats negative stereotypes. It has 115 chapters throughout the United States. The organization gives its members opportunities for professional development, networking, leadership training, and more. The group supports women builders and tradeswomen. Also, the government plays a role in promoting women through

There are many organizations that support and promote women in the construction field.

the Women's Apprenticeships and Nontraditional Occupations Act. This act gives federal grants to projects that support women in industries like construction.

The majority of building inspectors are white, at 68.5 percent. Hispanic or

Latino inspectors make up 13.1 percent, Black inspectors make up 8.4 percent, and Asian inspectors make up 3.4 percent. In the construction industry, 52.9 percent are white. About 28 percent are Hispanic or Latino, 11.1 percent are Black, and 2.8 percent are Asian.

"There is certainly a noticeable increase in female students," said Hofkin. "I have seen a steady number of Black and Hispanic students over the years." He said teens should definitely consider being an inspector. "For an inspector, the financial rewards are significant," Hofkin said. "The opportunities for career advancement are plentiful. The sense of satisfaction that you feel as you protect the public from **myriad** hazards that could result from improperly constructed buildings is indescribable. I love my job!"[7]

INSPECTIONS GO VIRTUAL DURING THE COVID-19 PANDEMIC

Some **municipalities** did building inspections virtually in 2020 at the start of the COVID-19 pandemic. Virtual inspections meant less spread of the virus. A contractor representative took videos and photos at the construction site. Then the representative sent the images to building inspectors online. Some towns have continued to use virtual inspections because it saves money.

GLOSSARY

fall arrest systems

systems that help keep people from falling

liability insurance

a type of insurance that protects people against personal injury and property damage claims

mallet

a wooden tool similar to a hammer

municipalities

towns or cities that have their own form of government

myriad

a great amount

provisional

temporary

rappelling

going down a building while being attached to a rope

stereotypes

often untrue beliefs that people have about other people or things

SOURCE NOTES

CHAPTER ONE: WHAT DOES A CONSTRUCTION AND BUILDING INSPECTOR DO?

1. Quoted in "La Rocca Orders Citywide Safety Sweep After Three Construction Workers Die," *Real Estate Weekly*, June 1, 2021. https://rew-online.com.

2. Pete Jackson, Personal interview, September 30, 2022.

3. Michal Hofkin, Personal interview, October 3, 2022.

CHAPTER TWO: WHAT TRAINING DO CONSTRUCTION AND BUILDING INSPECTORS NEED?

4. Michal Hofkin, Personal interview, October 3, 2022.

CHAPTER FOUR: WHAT IS THE FUTURE FOR CONSTRUCTION AND BUILDING INSPECTORS?

5. Quoted in Paul Lagasse, "ICC Chapters Are Cultivating Tomorrow's Workforce," *International Code Council*, October 14, 2019. www.iccsafe.org.

6. Michal Hofkin, Personal interview, October 3, 2022.

7. Michal Hofkin, Personal interview, October 3, 2022.

FOR FURTHER RESEARCH

BOOKS

Careers: The Ultimate Guide to Planning Your Future. New York: DK, 2022.

Sue Bradford Edwards, *Become a Construction Equipment Operator*. San Diego, CA: BrightPoint Press, 2023.

Stuart A. Kallen, *Skilled Jobs in Construction*. San Diego, CA: ReferencePoint Press, 2021.

INTERNET SOURCES

"Construction and Building Inspectors," *My Future*, n.d. www.myfuture.com.

"Occupational Outlook Handbook: Construction and Building Inspectors," *US Bureau of Labor Statistics (BLS)*, November 3, 2022. www.bls.gov.

"What Is a Construction Inspector?" *Zippia*, n.d. www.zippia.com.

WEBSITES

Build Your Future
www.byf.org

Build Your Future is a website where young people can learn about careers in construction. It gives information on how people can become a carpenter, electrician, plumber, project manager, welder, and more.

Building, Design, and Construction
www.bdcmagazine.com

Building, Design, and Construction is an online magazine about the construction industry.

The International Code Council
www.iccsafe.org

The International Code Council produces building codes that are used globally. It also offers classes toward certification for tradespeople and awards certificates to them.

INDEX

Brown, Jim, 62
building codes, 12–15, 17, 20–21, 23, 28–29, 36, 37–38, 40, 46, 64

Certified Building Official Limited Exam, 41
community colleges, 32–34
COVID-19 pandemic, 73

DeLuca, Amy, 9

Federal Emergency Management Agency (FEMA), 17
Flandro, Xsusha, 10

Hofkin, Michal, 28–30, 37–38, 44, 64, 66, 72–73

International Association of Electrical Inspectors, 57
International Code Council, 39, 62, 65

Jackson, Pete, 27, 36, 46, 55, 56, 64

La Rocca, Melanie, 26
licenses, 39, 41, 67

mentors, 35, 62, 64–65

National Association of Women in Construction, 70
National Electrical Code, 57
National Fire Protection Association, 39

permits, 20, 23
Pryor, Jason, 63–64

rope-access inspections, 8–9

salaries, 58–59
stop work orders, 22–23, 25

tools, 8, 53–55, 65
Turner, April, 6–9

UL Solutions Product Standards, 57

Women's Apprenticeships and Nontraditional Occupations Act, 71

zoning rules, 12, 18–19

IMAGE CREDITS

Cover: © KomootP/Shutterstock Images
5: © BearFotos/Shutterstock Images
7: © noomcpk/Shutterstock Images
9: © RainStar/iStockphoto
11: © RainStar/iStockphoto
13: © gnepphoto/Shutterstock Images
16: © strickke/iStockphoto
19: © YegoroV/Shutterstock Images
21: © MilanMarkovic78/Shutterstock Images
25: © T Photography/Shutterstock Images
26 (top left): © Chingraph/Shutterstock Images
26 (top right): © Motorama/Shutterstock Images
26 (bottom left): © kuroksta/Shutterstock Images
26 (bottom right): © Cube29/Shutterstock Images
29: © Zivica Kerkez/Shutterstock Images
31: © Alona Siniehina/Shutterstock Images
33: © vladgphoto/Shutterstock Images
35: © Chun Han/iStockphoto
37: © Andrey_Popov/Shutterstock Images
40: © Chris Ryan/iStockphoto
43: © Dragana Gordic/Shutterstock Images
46: © Andrey_Popov/Shutterstock Images
49: © SeventyFour/Shutterstock Images
50: © Viacheslav Nikolaenko/Shutterstock Images
52: © grandriver/iStockphoto
54: © Tongpool Piasupun/Shutterstock Images
56: © Electrical Engineer/Shutterstock Images
59: © Marek Masik/Shutterstock Images
61: © Olena Yakobchuk/Shutterstock Images
63: © Lisa F. Young/Shutterstock Images
66: © Amorn Suriyan/iStockphoto
68: © Manop Boonpeng/Shutterstock Images
71: © Manop Boonpeng/Shutterstock Images

ABOUT THE AUTHOR

Elizabeth Hobbs Voss is a children's author and journalist. She enjoys writing about subjects that are important to young people. With this book, she hopes to help teenagers decide whether they might like to pursue a career as a construction and building inspector.